For my family and yours.

We are all in the same boat.
It is time to row together.

We can make a difference.
Green Eco Warriors unite!

"Plastics, like diamonds, are forever!"

Captain Charles Moore, Algalita Marine Research Foundation

Pesky Plastic: An Environmental Story
© 2013 – Leticia Colon de Mejias
Printed in United States of America (USA)
Published by Great Books 4 Kids
(http://www.greatbooks4kids.org)

ISBN 978-0-9893364-1-3

Sally the sea turtle takes a swim in the sea.

She is looking for breakfast.

Look Sally!

What is that above you?

Is it a jellyfish for your breakfast?

Fast Fact: For millions of years everything in the ocean was organic matter that could be eaten by animals.

Oh no!

Wait!

Sally, don't eat that.

It is a plastic bag!

Plastic can look like a jelly fish to a sea turtle.

Fast Fact: *Each year people in America throw out enough plastic cups, spoons and forks to circle the earth's equator about 300 times.*

Sally does not know that eating plastic will make her sick.

Animals cannot tell the difference between plastic and food.

Sally the sea turtle and her friends live in oceans and on coastal beaches.

They depend on people to keep the ocean and the rivers clean.

Fast Fact: *Pollution from rivers and streams flows into the oceans. Cleaning up trash around our rivers and streams helps to reduce the amount of pollution that flows into the oceans.*

Pat the pelican flies low over the ocean.

Pat is looking for lunch.

What is that Pat sees in the water?

Is it a fish?

Fast Fact: *Americans use about one billion (1,000,000,000) plastic shopping bags each year. This creates 300,000 tons of waste and pollution.*

Oh No!

Stop Pat!

That is not a fish.

That is a plastic bag.

Pat must be careful.

Plastic can look like a fish to a pelican.

Fast Fact: *After you throw away plastic bags they do not disappear. Plastic is forever. Plastic is not biodegradable. Sunlight breaks plastic into tiny pieces, and these pieces float in the ocean or they are eaten by animals.*

Pat needs to find fish.

The ocean has strong currents that carry trash to far-away places.

Even on beaches where there are no people, there is plastic trash.

Trash that starts out in California may end up in Hawaii.

Fast Fact: *People throw away a lot of plastic things like bags, baby bottles, toys, toothbrushes, furniture and other items. Some of this plastic ends up in our rivers and oceans.*

Pat must find food to survive.

If Pat finds more plastic than fish,
then Pat may eat plastic by mistake just like
Sally the sea turtle.

Fast Fact: *Each time you recycle, you are removing trash from the environment. By recycling plastic bags and bottles, or choosing not to use throw-away items, you and your family are making our oceans and the planet healthier for animals and humans.*

Allen the albatross waits in his seaside nest
for his mother to feed him.

What is that Allen?

Is it a snack on the beach?

Fast Fact: *Every year, Americans throw out more than 7 billion pounds of plastic, but sadly we recycle less than 1% of the plastic we throw away.*

Birds that live near coastlines or oceans see more plastic than fish or plankton. In the Pacific Gyre, plastic now outnumbers plankton 10 to one.

No Allen. Stop!

That is a plastic lighter!

Plastic is not a snack.

Garbage can look like food
to a baby albatross.

Fast Fact: *Mother birds ingest plastic, thinking it is food. The babies are fed the ingested plastic through regurgitation. Many baby albatrosses die this way.*

Allen lives in a nest next to the sea.

He needs a place to live without garbage everywhere.

His mother needs to find food to feed him.

Fast Fact: Most grocery stores and shopping centers offer tote bags and reusable shopping bags for sale. Make an effort to use fabric or paper bags. You could even ask your parents to help you make your own reusable tote bag!

We can all help save the environment by using less plastic and being more aware of the effect we have on our surroundings.

Fast Facts: *Each time you go shopping, you have a chance to buy items that are not packaged or bagged with plastic. If we all stand together and use our buying power, we can make companies change their packaging to be more environmentally friendly.*

Plastic seems easy to use and throw away,
but sometimes the easy choice is not
the right choice!

By finding ways to avoid using plastic,
we can reduce the pollution it causes.

Everyone can make a difference.

Fast Fact: *Most of the plastic we use is produced far away from where we live and must be transported across the country or the oceans. By reducing, reusing and recycling, we also help to reduce the pollution created by the number of trucks and large ships needed to transport plastic.*

What can you do to make a difference?

Our Planet's oceans are all connected. This means that when you help in your neighborhood, you are helping the world. Help stop Pesky Plastic and become **Green Eco Warriors**.

Which of these items will help you avoid putting more plastic into the environment?

Plastic Bag

Foam Tray

Lunch Bag

Plastic Wrap

Paper Bag

Reusable Tupperware

Recycle Bin

Tote Bag

Answer: Lunch Bag, Paper Bag, Reusable Tupperware, Recycle Bin and Tote Bag.

How to make a Fabric Bag

Have an adult help you measure, cut and sew the fabric. You can use old fabric like T-shirts or jeans or canvas.

1. Measure and mark the fabric pieces.
2. Cut the fabric pieces to size.
3. Sew (or use fabric glue) the pieces together.
4. Color a design on your bag with fabric paint.
5. Use your bag when you shop . It will help you avoid using plastic bags.

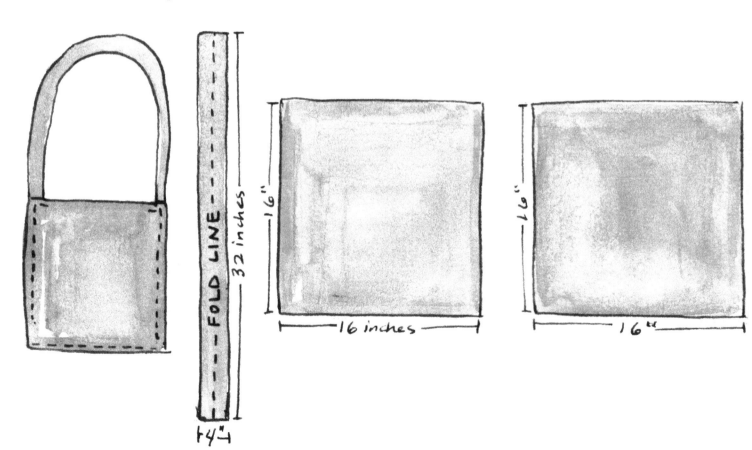

Teacher and Family Educational Guide

Although plastic items like bottles and bags are very easy to use, they come with a cost. Like diamonds, plastic is forever. When plastic ends up in our rivers or oceans, it can entangle and hurt or kill animals like sea turtles, seals and dolphins. When plastic breaks down into smaller pieces or particles, it looks like food and gets eaten by fish and seabirds. When an animal's stomach has plastic in it, the animal doesn't have room for food and water and could starve.

1. What is most plastic made from?

Plastics are made from petroleum (oil) that is dug up from deep underground.

Even before plastic becomes throw-away trash it does environmental harm. The process of making plastic requires large amounts of heat and energy. It releases chemical by-products and toxic vapors, that pollute the air and water, harming animals and people.

2. What are three ways that plastic litter can hurt ocean animals?

1. When plastic ends up as litter in the ocean, it can catch or entangle animals.
2. Plastic can look like food. According to the National Wildlife Foundation, more than one million (1,000,000) animals die every year by eating or getting tangled in plastic trash.
3. Plastic releases or leaches toxic chemicals into the water that are known to cause health issues in animals and humans.

3. How can we reduce the amount of plastic in the environment?

We can use fewer disposable plastic items such as; water bottles, plates, forks, bags and plastic wrap. We can become *Green Eco Warriors* and use reusable water bottles, fabric or paper shopping bags, reusable sandwich containers, and plates that we can wash instead of throwing them away.

We can avoid buying items that are overpackaged and have extra wrapping or plastic, especially packaging that is not recyclable.

We can recycle, which helps trash stay out of the environment. However, avoiding disposable plastic products to begin with is the most effective measure.

We can write our local and state representatives, along with our country's leaders about our concerns about plastic. We can ask our leaders to pass laws that protect people and the planet.

Plastic and People

Most plastic is made with unhealthy chemicals. These harmful or toxic chemicals are used to mold plastic into shapes like water bottles. When we drink from plastic bottles or eat food served in plastic containers, we sometimes ingest a small amount of these chemicals. This is especially true if we heat plastic. For example, food or beverages warmed in a microwave oven that are wrapped in plastic or served in plastic containers will release higher levels of the toxic chemicals into the food or drink. These chemicals have been shown to affect reproduction and brain development in animal studies, as well as cause cancer and other health problems. Instead use containers and bottles made from steel or glass.

Plastic, Fish and Plankton

In parts of our oceans, there is more plastic than plankton or fish. Over time with consistent exposure to the sun and water, plastic breaks down into lots of little pieces. Fish mistake the plastic for food and ingest the pieces. For billions of years it was safe for fish and other animals to eat objects floating in the water, this is because anything in the ocean was organic matter. Today sea animals are dying because they are eating plastic instead of organic matter.

Although we don't often think about plankton, these little guys play a big role in the planet's eco-system. They are a food source for many species of fish and whales. In areas of the ocean where there are large amounts of plastic, the amount of plankton has decreased to alarming levels. According to National Geographic the phytoplankton in our oceans removes 50% of the carbon dioxide in the planet's atmosphere, and creates 50% of the oxygen in air we breath. If the plankton in our oceans continues to disappear, so will the air we breath.

Polystyrene (also known as Styrofoam)

Polystyrene is one of the most widely used petroleum-based plastic. Polystyrene products come in many shapes and sizes like Styrofoam cups, trays, packaging peanuts and containers. In the United States alone, we throw away about 25 billion Styrofoam coffee cups every year. The Environmental Protection Agency (EPA) named the polystyrene manufacturing process as the fifth largest creator of hazardous waste. The National Bureau of Standards Center for Fire Research identified 57 chemical by-products released during the combustion of polystyrene foam. The process of making polystyrene pollutes the air and creates large amounts of liquid and solid waste. Toxic chemicals leach out of these products into the food that they contain (especially when heated in a microwave). These chemicals threaten human health, including reproductive systems.

Glossary

Biodegradable - capable of being broken down by bacteria or other living organisms.

Chemical - a compound or substance purified or prepared that does not normally occur in nature. It is produced by or used to alter the atomic and molecular composition and structure of the substances involved.

Coastal beaches - usually consists of loose particles that are often composed of types of rock such as sand, gravel, shingle, pebbles or cobblestones where the land meets the ocean or sea. It also consists of shells and coralline algae which are biological in origin.

Ingest - to take food into the body through the mouth; to eat.

Gyre - a gyre is a naturally occurring mass of whirling ocean current and air that rotates in a clockwise direction in the Northern Hemisphere and counterclockwise in the Southern Hemisphere. The whirling current and air create a whirlpool effect that moves more slowly at the center . The slow moving circle is where plastic debris collects. For example, the North Pacific Gyre, located in the northern Pacific Ocean, is one of the five major oceanic gyres. This gyre comprises most of the northern Pacific Ocean. It is the largest ecosystem on our planet.

Ocean currents - the steady flow of ocean water in one direction.

Organic matter - is matter composed of organic compounds from the remains of once-living organisms such as plants and animals and their waste products in the environment. Organic matter is very important in the movement of nutrients in the environment and plays a role in water retention on the surface of the planet.

Plastic - a man made material made from a wide range of polymers such as polyethylene, PVC and nylon that can be molded into shape while soft and then set into a rigid or slightly elastic form by chemical processes.

Plankton - small and microscopic plants and organisms drifting or floating in the ocean, sea or freshwater, consisting of algae, one-celled living organisms, small crustaceans, and the eggs and larval stages of larger animals. Many animals feed on plankton.

Regurgitation - to pour back or cast up partly digested food (as is done by some birds in the feeding of their young).

Toxic - poisonous or containing poisonous material capable of causing death or serious debilitation.

Teacher Guide - Classroom Waste Analysis Activity

Items needed for each team:
- Camera
- Notebook
- Cardboard box or trash bin

STEPS:
- Create teams of two or more students, each team should have a worksheet, pencil and trash bin or cardboard box.
- Take your class outside to collect trash. Place the collected trash in the trash bin or box.
- Count and categorize the trash in two ways:
  1. Trash that is biodegradable vs. plastic and other non-biodegradable trash
  2. Recyclable trash vs. non-recyclable trash

Take photos of your collected trash pile and of your class recycling them. Send us the photos and the categorized counts. We may post your class on the website or feature them in the next issue of the Green Eco Warriors National Publication.

Discussion

How has trash changed? What are the differences in the amount and types of trash created today as opposed to trash in the past?

Like diamonds, plastic is forever. It has an endless life expectancy and only breaks into smaller toxic pieces. When burned it releases toxins, and when left in water or buried, it leaches toxins such as BPA into the ocean or drinking water. Before plastic was used in everyday throw-away applications, glass, canvas, and tin were commonly used. Lunches were wrapped in wax paper and carried in metal lunch boxes. Groceries were packed in canvas sacks. These items lasted longer and were commonly reused instead of being thrown away like plastic sandwich bags or plastic shopping bags.

Why Recycle?

By recycling, you will be protecting the environment, protecting your health and the health of the people you love.

Fast Facts on Recycling

- Reduces the amount of waste that must be disposed of – which means less waste to incinerate (burn) which creates air pollution and greenhouse gases, - or dumped in a landfill which can lead to groundwater pollution.
- Conserves water and precious natural resources – since less natural resources need to be extracted from the earth and processed.
- Saves energy – In 2003, EPA reported the energy savings from recycling in the U.S. accounted for roughly 1,486 trillion Btu in energy savings - an amount equivalent to the consumption of 11.9 billion gallons of gasoline or 256 million barrels of crude oil.
- By using less natural resources and using less energy we reduce greenhouse gas emissions.

The energy saved by recycling also results in less pollution and we all know how bad pollution is to our environment. When you make new products from used materials precious natural resources are saved.

How to Start Recycling

Recycling is easy and important!

1. Learn what to recycle in your area by searching for your town or city name, state and the words "recyclable items" on the Internet.
2. Think about it, could the items you are throwing in the trash be recycled?
3. Collect the items that are recyclable inside a paper bag or recycling bin. You can leave the items curbside with your trash for pickup on trash day.
4. If your city or town does not offer curbside recycling, work with them to start a recycling program. Invite your family, friends, neighbors, schools, local businesses, city and state elected officials to support one.

Schools may contact the U.S. Environmental Protection Agency (EPA) at http://ww.epa.gov to learn more about coordinating large recycling pickups on trash days.

Thank you for your interest in living sustainably.
We can all be Green Eco Warriors.
Education and Motivation, Fight for Our Planet's Survival.

About the Author

Leticia Colon de Mejias lives in Connecticut with her husband, five children, and their dogs, Bio-diesel and Lady Bird. Leticia's motivation for doing all she can to help the environment comes from her love of children and nature. She has authored and illustrated seven children's books including "Butterfly Rhythm", "Hip Hop and the Wall", "Mrs. Busy Bee", "Dinero the Frog Learns to Save Energy" and "Over the Moon and Past the Stars." She donates time and books to local schools and libraries. After watching *Kilowatt Ours* and *Message in the Waves*, she co-founded the Green Eco Warriors.

Energy Efficiencies Solutions (EES)

EES is a Connecticut-based energy conservation company. Invested in conservation and youth education since 2008, EES has partnered with schools, libraries and the Green Eco Warriors to provide free education and programs focused on recycling, energy conservation, and empowering youths to make a difference and to fight for our planet's survival. Visit us at *www.EESgoGreen.com* to learn more.

About the Illustrator

Drawing and painting have been part of Tamara Visco's life since childhood. She took art classes in school, but she also practiced a lot on her own and learned from family and friends along the way. Tamara has done several children's books including "Dinero the Frog Learns to Save Energy." From the start, she was enthusiastic about working on this project because working together to save the planet is so important for everyone.

Our Planet - Our Future

"Fighting to save our planet through education and motivation"

Become a Green Eco Warrior, save the planet and win prizes!

Join us at www.GreenEcoWarriors.org

Green Eco Warriors provides Sustainability Education aligned with national standards. Our lessons take a holistic approach to learning, incorporating environmental, health, and social perspectives in a format children can easily comprehend.

Green Eco Warriors Monthly Issues Contain:

- Monthly articles that explain the environment, energy and issues that have an effect on health and the planet.
- Discussion questions
- Writing exercises
- Extended learning activities including:
    1. Activities that engage students and parents
    2. Learning vocabulary
    3. Student research and presentation projects
    4. Prizes and events for challenge participants
    5. Opportunities to participate in local community service events

Teacher guide included in every issue

Lessons reinforce national education standards:

- Reading comprehension
- Critical thinking
- Writing Skills
- Vocabulary
- Listening and debating
- Research and presentation skills
- Current events and issues

- Geography
- Science
- STEM
- Common Core
- Next Generation Standards
- 21st Century Skills

To learn more about what you can do at home, in your school, and in your community, visit us at *www.GreenEcoWarriors.org*.

Other Great Books 4 Kids

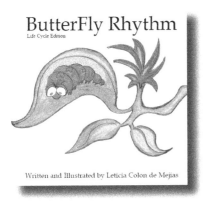

ButterFly Rhythm
Life Cycle Edition

Written and Illustrated by Leticia Colon de Mejias

Being little can be tough. You can't really see much from low to the ground. As Butter the caterpillar grows she finds she has a much larger view of the world. This is a story of a caterpillar that grows into a butterfly she did not know existed.

Book includes the life cycle of a butterfly, science lesson appropriate for grades K - 2.

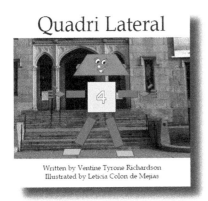

Quadri Lateral

Written by Ventine Tyrone Richardson
Illustrated by Leticia Colon de Mejias

Join Quadri Lateral, the boy made of four sided shapes, in his exploration of "Polygon City." Meet his dog "Tripod" and follow the two friends as they explore the many shapes that make up the polygon family. Quadri will help kids learn to identify polygons in the Hartford, CT, skyline. He demonstrates that you can find polygons and other shapes in real life applications. Quadri introduces easy ways to remember what makes a polygon a polygon. The illustrations and city photos will draw even the most skeptical child into the process of learning about polygons and quadrilaterals.

DINERO THE FROG
LEARNS TO SAVE ENERGY

Written by Leticia Colon de Mejias
Illustrated by Tamara Visco

Dinero the Frog Learns to Save Energy is a fun and educational book about energy conservation. Poppi the Frog teaches Dinero about energy, where it comes from, how it is used, and what we can do to conserve energy and reduce pollution.

Book includes questions and answer activities as well as a glossary of frequently used energy conservation terms appropriate for grades K - 5.

CPSIA information can be obtained
at www.ICGtesting.com
Printed in the USA
BVHW01s1528230318
511248BV00010B/103/P